# Laboratory Management
## Quality in Laboratory Diagnosis

*Diagnostic Standards of Care*

MICHAEL LAPOSATA, MD, PHD
*Series Editor*

## Coagulation Disorders
Quality in Laboratory Diagnosis
*Michael Laposata, MD, PhD*

## Clinical Microbiology
Quality in Laboratory Diagnosis
*Charles W. Stratton, MD*

## Laboratory Management
Quality in Laboratory Diagnosis
*Candis A. Kinkus, MBA*

*Forthcoming in the Series*
## Clinical Chemistry
## Transfusion Medicine
## Hematology / Immunology

*Diagnostic Standards of Care Series*

# Laboratory Management
## Quality in Laboratory Diagnosis

**Candis A. Kinkus, MBA**

Administrative Director
Diagnostic Laboratories
Vanderbilt University Medical Center
Nashville, Tennessee

**demos**MEDICAL
New York

Visit our website at www.demosmedpub.com

*ISBN*: 978-1-936287-45-1
*E-book ISBN*: 978-1-617050-88-6

*Acquisitions Editor*: Richard Winters
*Compositor*: S4Carlisle Publishing Services

**Library of Congress Cataloging-in-Publication Data**

Kinkus, Candis A.
    Laboratory management : quality in laboratory diagnosis / Candis A. Kinkus. p. ; cm.—
(Diagnostic standards of care series)
    Includes index.
    ISBN-13: 978-1-936287-45-1
    ISBN-10: 1-936287-45-5
    ISBN-13: 978-1-61705-088-6 (e-book)
I. Title. II. Series: Diagnostic standards of care.
    [DNLM: 1. Laboratories—organization & administration. 2. Total Quality Management. QY 23]
616.02'7—dc23                                                    2011038669

Printed in the United States of America by Gasch Printing
11  12  13  14/ 5  4  3  2  1

*To my husband, Charles Fulwider, and children, Tim and Carolyn, for their unwavering support and encouragement*

# Contents

# Series Foreword

"Above all, do no harm." This frequently quoted admonition to health care providers is highly regarded, but despite that, there are few books, if any, that focus primarily on how to avoid harming patients by learning from the mistakes of others.

Would it not be of great benefit to patients if all health care providers were aware of the thrombotic consequences from heparin-induced thrombocytopenia before a patient's leg is amputated? The clinically significant, often lethal, thrombotic events that occur in patients who develop heparin-induced thrombocytopenia would be greatly diminished if all health care providers appropriately monitored platelet counts in patients being treated with intravenous unfractionated heparin.

It was a desire to learn from the mistakes of others that led to the concept for this series of books on diagnostic standards of care. As the test menu in the clinical laboratory has enlarged in size and complexity, errors in selection of tests and errors in the interpretation of test results have become commonplace, and these mistakes can result in poor patient outcomes. This series of books on diagnostic standards of care in coagulation, microbiology, transfusion medicine, hematology, clinical chemistry, immunology, and laboratory management are all organized in a similar fashion. Clinical errors, and accompanying cases to illustrate each error, are presented within all of the chapters in several discrete categories: errors in test selection, errors in result interpretation, other errors, and diagnostic controversies. Each chapter concludes with a summary list of the standards of care. The most common errors made by thousands of health care providers daily are the ones that have been selected for presentation in this series of books.

Practicing physicians ordering tests with which they are less familiar would benefit significantly by learning of the potential errors associated with ordering such tests and errors associated with interpreting an infrequently encountered test result. Medical trainees who are gaining clinical experience would benefit significantly by first understanding what not to do when it comes to ordering laboratory tests and interpreting test results from the clinical laboratory. Individuals working in the clinical laboratory would also benefit by learning of the common mistakes made by health care providers so that they are better able to provide helpful advice that would avert the damaging consequences of an error. Finally, laboratory managers and hospital administrators would benefit by having knowledge of test ordering mistakes to improve the efficiency of the clinical laboratory and avoid the cost of performing unnecessary tests.

If the errors described in this series of books could be greatly reduced, the savings to the health care system and the improvement in patient outcomes would be dramatic.

*Michael Laposata, MD, PhD*
*Series Editor*

# Preface

A clinical laboratory service is a complex operation. It offers a vast array of tests, performs these tests using diverse methodologies and then provides accurate and timely results, which are often used to make decisions in life and death situations. Managing a clinical laboratory, not surprisingly, is a complex challenge. In this volume of the series, Diagnostic Standards of Care, an instructive case-study approach provides numerous examples of laboratory management problems and the ways in which they can be averted.

This book examines the broad scope of responsibilities that encompass laboratory management. The case study format illustrates decisions where the outcomes produced negative, and sometimes grave, consequences for patient care. These case studies are based on actual incidents that have occurred in the workplace, from community hospital settings to academic medical centers. As with the cases in all volumes of this series, they present an opportunity to learn from the mistakes of others.

The topics in this volume are organized across a dozen different areas of management responsibilities. These topics are diverse, ranging from staff management to regulatory compliance and from market competition to technology infrastructure. A common theme demonstrated through many of these case studies is the benefit of collaboration with others to resolve a problem. Many of these case studies illustrate that a team approach to management, which engages key stakeholders, provides broader insight and can lower the risk of errors and problems.

Laboratory professionals who are moving into management roles are encouraged to broaden their knowledge base beyond technical expertise, through formal and informal programs. It is essential to develop management skills.

*Candis A. Kinkus, MBA*

# Acknowledgments

I would like to acknowledge Michael Laposata, MD, PhD, the series editor, for his excellent insight and direction in the development and focus of this volume. In addition, I greatly appreciate the patient guidance and instruction of Richard Winters, Executive Editor at Demos.

# Accreditation and Regulatory Compliance

## OVERVIEW

Laboratory services are regulated at the federal level, and in many cases, by one or more state agencies as well. If state regulations are more stringent than federal, then the state regulations supersede federal. In addition, there are federally designated, nongovernment agencies that conduct periodic and unannounced on-site inspections of laboratories. These inspections serve to document performance compliance with standards and thereby accredit the laboratory to receive or maintain a federal license to operate. This federal license is referred to as a "CLIA" license as laboratory licensure was deemed a regulatory requirement with the enactment of the Clinical Laboratory Improvement Act (CLIA) of 1988.

Good management requires that the leadership team understand both the federal and state laws as well as the accreditation standards established by the inspecting agencies. It is the leadership team's responsibility to assure that actual laboratory practice complies with both legal regulations and accreditation requirements. Failure to do so may lead to consequences ranging from almost nothing to the catastrophic for the laboratory service: for example, from a substandard performance that requires a repeat inspection to unacceptable performance with suspension of license and testing.

# UNDERSTAND REQUIREMENTS OF STATE REGULATIONS

▶ The management team must be aware of government (federal, state, and local) regulations, accreditation standards, and institutional policies. These performance requirements apply to a broad range of laboratory activities, including personnel, safety, reports and records, claims billing, and waste disposal.

### Case with Error

The institution's Human Resources Department has a standard policy of a minimum of 5 years full-time experience for all supervisor positions. In this state, regulations require a minimum of 6 years full-time laboratory experience for supervisory positions.

An inspector was conducting an accreditation inspection that included a review of job descriptions. She noted the discrepancy for experience in the supervisor job description and requested a review of all supervisory personnel records. All personnel records indicated that the currently employed supervisors met state requirements.

### Explanation and Consequences

Laboratory management is responsible for assuring that the job descriptions meet the human resources policies, laboratory regulations, and accreditation standards. Neither federal law nor accreditation standards specify any requirements for experience in a supervisory position. Since this state's regulations define a minimum standard, it is the most stringent requirement and must be met by all laboratories in that state. Laboratory management failed to verify the state law and in this instance, did receive a citation with a requirement to modify the job descriptions. When there are supervisory employees who do not meet the minimum years of experience, the hospital could incur fines and, most likely, both laboratory operations would be disrupted and employees impacted with removal of supervisors who do not meet the standard.

### Case with Error Averted

A laboratory's record retention policy defined that documents should be maintained for 2 years, as per accreditation standards. This policy applied to various records such as quality control logs, proficiency test results, quality management (QM) reports, and so on.

In the state where this laboratory operated, the legislature modified the laws pertaining to medical malpractice. These revised laws permitted plaintiffs to file a civil suit for a period of time up to 7 years from the date of incident. It also allowed the plaintiff to subpoena records for 3 years prior to the date of incident.

At the direction of its legal counsel, the laboratory revised its record retention policy to a period of 10 years.

### Explanation and Consequences

Since the revised state law is more stringent, it supersedes the accreditation standard. In the event of any legal action, the laboratory must provide any necessary documentation as proof of performance. If the laboratory failed to maintain these records as legally required, it would place the institution at significant risk in a lawsuit where test results were relevant to the case.

# PREPARE FOR INSPECTION

▶ Unannounced inspection for licensure is a routine practice. At a moment's notice, the laboratory management team must produce numerous test procedures and quality report records that demonstrate performance over a 2-year period. This requires a standard practice for documenting and maintaining records of test and instrument performance, quality control, written procedures, and reviews, so that these can be readily produced at the time of inspection.

### Case with Errors

Over the course of a full-day inspection, the laboratory supervisor failed to produce a number of routinely requested records. Records were disorganized and there were "gaps" where some records were sporadically missing for various periods of time. For those records that were produced, these documents demonstrated acceptable performance.

During an inspection, failure to produce the necessary documents is essentially interpreted as failure to perform the work. As a result, the laboratory received 50 citations for primary deficiencies and 12 citations for secondary deficiencies.

### Explanation and Consequences

The laboratory management team did not establish a standard process for documenting and maintaining records, which could be followed by all staff. These failures were the contributory factors to the numerous citations. Although the laboratory maintained licensure, the high number of primary deficiencies required a second on-site inspection within 6 months.

# COLLABORATE WITH INSTITUTIONAL DEPARTMENTS

▶ There are accreditation standards and regulations that define key data elements that must appear on a test results report. However, the laboratory test report is also a medicolegal record. As such, all test results and comments are subject to interpretation in any legal proceeding.

## Case with Errors

Several years of patient test results were converted from the originating laboratory information system (LIS) database into a new vendor's LIS software. The laboratory leadership team wrote a simple disclaimer statement to appear on every converted result report that stated that the results were not from the source database and may not be complete. At a later date, a patient filed a civil suit against the hospital and the disclaimer statement on the converted result reports unexpectedly became a point of contention.

## Explanation and Consequences

While the disclaimer statement was a simple statement of fact, the language was not as clear and precise as required for a legal document. It is important to understand that all information contained on a laboratory result report should be reviewed through the institution's established process for vetting and approving all forms that comprise the patient's legal record. If the legal department is not part of this process, it is recommended to include it.

# STANDARDS OF PERFORMANCE

▓ Laboratory management should verify and understand regulatory requirements at all levels of government: federal, state, and local. Federal regulations, including CLIA, routinely state that the federal requirements are a minimum performance standard and do not supersede more stringent standards from state and local government agencies. It is also important to know the performance standards for any other regulatory agency, such as College of American Pathologists, the Joint Commission, and American Association of Blood Banks, that have accreditation oversight of the laboratory. The laboratory's procedures and policies must meet the most stringent requirements imposed by the applicable regulatory entity or accrediting agency.

▓ While laboratory inspectors arrive unannounced, they are fully prepared and expect to complete all aspects of inspection within a specific time frame for that site. It is imperative that the management team has defined a standard process for maintaining necessary and complete documents and records, so that these can be readily produced at the time of inspection.

▓ The laboratory test result report is a medicolegal document and this applies to all information contained on the report, not just the test results. Statements that are simple and clear in everyday communication may not meet those legal standards in a court of law. As test result reports are created or revised, the management team is responsible for assuring that these documents are properly reviewed and approved.

# 2  Patient Safety

## OVERVIEW

The Institute of Medicine report published in 2000 clearly documented an unacceptable number of negative outcomes for patients. One of the key factors contributing to the problems in patient care is the failure of health care providers to define safe practice standards and consistently enforce compliance. This report led to the development of the patient safety standards by numerous agencies, including The Joint Commission and the College of American Pathologists.

A key patient safety standard for the laboratory is to ensure positive patient identification (PPID). It is a required practice that PPID must use two unique patient identifiers. These identifiers usually include the patient's complete name and one or more of the following: medical record number, complete birth date, or Social Security number. Once a specimen has been received in the laboratory and is in process, a common practice is to assign a unique accession number that may be used in conjunction with the patient's name during the testing process.

The PPID practice is a critical step in the preanalytic phase of laboratory testing. Some accrediting agencies require that at the time of specimen collection, health care providers must use both "active and passive" identification methods. Passive identification requires the health care provider to verify the printed patient identifiers (on specimen labels, test requisitions, or computer screen) with a patient identification armband that must be attached to the wrist or ankle of an admitted patient. For active identification, the employee is required to engage the patient to verbally confirm his or her identity. A common practice is to request the patient to state his or her complete name, spell his or her last name, and provide his or her complete date of birth. The employee must verify that the patient's responses match the printed identifiers (on specimen labels, test requisitions, or computer screen).

During the analytic phase, many test methods require manual processes. It is necessary to design workflow to ensure that the patient identity is correctly maintained with the specimen throughout testing.

Likewise, in the postanalytic phase, there are some circumstances that may warrant providing a verbal result to the clinician. It is imperative that this communication between both parties occurs with the use of PPID.

Regardless of the accuracy of the test method, a correct result provided for the wrong patient can have dire consequences, including loss of life. Laboratory leadership is responsible for defining clear procedures for patient safety, training staff to assure competency, and consistently maintaining accountability for compliance by all staff at all times.

# PERFORM PROPER SPECIMEN COLLECTION

▶ When performing PPID, the phlebotomist often has multiple computer-generated labels that will be applied to each test tube. It is necessary to verify the patient identifiers on each printed label, not just on the first label.

## Case with Error

Two patients, a woman and her daughter-in-law, were admitted on the same day and to the same semiprivate room, but for different procedures. The phlebotomist arrived at the patients' room and had six preprinted specimen labels. She performed both active and passive identification with the daughter-in-law, collected all six tubes of blood, and applied the labels onto the tubes at the patient's side. However, the phlebotomist read only the first one of the six labels. The last two labels were for tests ordered for the mother-in-law.

The phlebotomist's failure to read and verify each printed label meant that the last two tubes were mislabeled and contained the wrong blood. Those tests ordered for the mother-in-law were performed on blood collected from the daughter-in-law.

## Explanation and Consequences

This error was identified by the laboratory staff when the blood type and antibody screening test ordered for the mother-in-law was performed and the blood type did not match the historical blood type of the mother-in-law. This discrepancy in the blood type results triggered an incident investigation where a blood type was performed on each of the six test tubes collected by the initial phlebotomist. The results demonstrated that the same blood type result was produced for each test tube. In addition, a new specimen was collected by a different phlebotomist from each patient and tested for blood type. Results from the second specimen collection confirmed the historical blood type for the mother-in-law and the "new" blood type for the daughter-in-law.

When collecting specimens where there are multiple labels printed with the patient identifiers, it is necessary to verify each printed label. A common practice with computer-generated labels is to program the software to allow for a blank label between each patient. This blank label can then serve as a visual cue to the phlebotomist that these are labels for a different patient. It must be emphasized that even with the use of blank labels, it is still required practice to verify the patient identifiers on each label.

This laboratory's written procedure did specify that each printed label must be verified when performing active and passive identification at the time of specimen collection. Thus, this incident was caused by the employee's failure to comply with the defined procedure.

▶ In the outpatient setting, patients present for specimen collection, but do not receive a patient identification armband. However, the patient does have a physician's written test order with the patient's complete name. A common practice is to use this document in lieu of the identification band. When performing passive and active patient identification, it is incumbent upon the phlebotomist to verify all computer-generated labels and the patient's verbal identification with the written test order.

### Case with Error

At a phlebotomy site for outpatients, a female patient presented with a written test requisition for Cara Corcoran and a birth date of April 12, 1990. The clerk accessed the registration database and selected an existing registration for Carly Corcoran with the birth date of April 12, 1990. The laboratory clerk then entered the test orders and printed specimen labels with the name Carly Corcoran.

The phlebotomist obtained both the labels and the requisition and called the patient name listed on the test requisition, Cara Corcoran. Prior to venipuncture, the phlebotomist performed an active patient identification only with the requisition which did match the patient's verbal statement of her name, Cara Corcoran, and birth date. The phlebotomist collected the specimens and then applied the preprinted labels to the tubes at the patient's side. These tubes were delivered to the laboratory a short time later. Upon receipt in the laboratory, the laboratory assistant compared the labeled specimens with the requisition and identified the discrepancy.

### Explanation and Consequences

This patient, Cara Corcoran, is one of a set of triplets. Her siblings are named Carly and Cassie. Upon investigation of this incident, the clerk admitted that she noticed that the first names were different yet similar and assumed there was a typographical error in the registration database. The phlebotomist stated that she did not verify the patient identification on the labels at any point during the venipuncture procedure.

In this example, the clerk identified a discrepancy in the first names but since the birth dates matched, she assumed the person registered in the database was the same person presenting for laboratory tests. The phlebotomist did not comply with one of the performance requirements as she did not confirm the information on the labels. It was only at the third step in the preanalytic process that the laboratory assistant identified the discrepancy with the first names. These employees' errors did require that the patient had to return and submit to a second venipuncture.

The patient safety standard clearly states that verification of patient identification must be performed with the patient's complete first and last name. This facility designed its process so that a different individual was responsible for performing each preanalytic step: test order, venipuncture, and specimen receipt. By allocating these tasks across numerous individuals, it provides redundancy so that if one employee fails to complete all verification tasks, a subsequent employee will do so and thus increase the probability of detecting an error.

> ▶ In addition to performing both passive and active patient identification, another requirement of PPID is to label all specimen containers at the patient's side. This practice minimizes the risk of error that a label could inadvertently be applied to a specimen collected from another patient.

### Case with Error

A 35-year-old mother (Patient A) of four was admitted for a surgical procedure. The phlebotomist presented at the patient's bedside, properly performed PPID, and collected specimens for all ordered tests. He left the patient's room with both the unlabeled specimens and the preprinted labels and proceeded to the workstation. He then placed the tubes on the counter and applied labels. A colleague was working at the same counter and was also labeling several specimens collected from another patient (Patient B).

All tubes were delivered to the laboratory and testing was completed, which included a blood type and cross match. This female Patient A did not have a previous history of blood type at this hospital. The following day, several units were transfused intraoperatively to Patient A. She had a massive transfusion reaction and expired.

### Explanation and Consequences

This failure to comply with procedure produced a catastrophic outcome. A thorough investigation of this sentinel event was conducted. It was found that the blood type for all other test tubes labeled with the expired Patient A's name was different from the blood type from the test tube used for the type and cross-match test. The phlebotomist admitted that he did not label specimens at the patient's bedside and also stated that a coworker was labeling another patient's specimens at the same workstation. Those specimens collected from Patient B and labeled by the coworker were also tested for blood type. It was found that one of the tubes was a different blood type from all of the other tubes labeled for Patient B and that this different blood type matched Patient A's blood type.

It was determined that during the specimen labeling process, one unlabeled specimen from each patient was somehow "swapped" at the work counter and mislabeled. It is imperative that all staff understand the importance to comply with the defined process for collecting and labeling specimens. Failure to do so will introduce an opportunity for error and place the patient at risk, as clearly demonstrated in this incident.

## VERIFY PATIENT IDENTIFICATION IN THE ANALYTIC PHASE

Many test methods require that an aliquot of the specimen is manually transferred from the original specimen container to another container (tube, well, plate, cassette) for testing. It is absolutely necessary to verify the patient identification on the original specimen container with the patient identification on the test container. Standard procedure is to perform this specimen transfer process on a one-by-one basis. The employee should only work with one patient at a time when transferring specimens from one container to another. This practice reduces the risk of erroneously transferring a sample from one patient to a container identified for another patient.

To minimize the amount of blood collected from a patient, it is common practice to use one tube of blood or body fluid for tests that are performed in multiple departments. The usual practice is to properly label new test containers and then aliquot samples from the original specimen. Patient identifiers on the original tube and aliquot tubes should be verified as the samples are dispensed.

If the initial test result is positive and the method requires a repeat test to validate an original positive result, it is recommended to obtain the original specimen container, if possible, and perform the repeat test with a sample from the original specimen container.

### Case with Error

A patient who was a candidate for organ transplant underwent periodic testing for various tests, including HIV status. At one point, the patient's test results for HIV were positive. The clinician contacted the laboratory and was concerned that this patient had not engaged in any activity that would produce a positive HIV test result that would alter his status as a candidate for organ transplant. It was agreed to collect a new specimen and repeat the test, which was performed by a different technologist than the one who performed the test with a positive result. The test on the new specimen produced a negative result.

### Explanation and Consequences

This laboratory routinely received a specimen aliquot for HIV testing. When an initial result was positive, a repeat test was performed from the same specimen aliquot. The investigation confirmed that this process was followed. The technologist who performed the test with a positive result verified that she followed standard procedure for transferring the specimen from the aliquot container to the test container. It was verified that a specimen from another patient with a history of positive results was tested in the same run with the organ transplant patient's specimen. Unfortunately, none of the original specimen containers or aliquot containers were still available for further testing as part of this investigation.

The outcome of the investigation was to change the process and require that the repeat test must be performed from both the original specimen container and the aliquot container. In those situations when the results from the original specimen container and aliquot container are discrepant, no result would be reported. Instead, the laboratory will notify the clinician about the discrepant results and request a new specimen and repeat testing at no cost to the patient.

In this particular situation, this erroneous result certainly created undue anxiety for the patient and family. Fortunately, during the short period of time when the patient was technically "ineligible" for an organ transplant, a donor match was not available.

> The importance of performing specimen transfer on a one-by-one basis cannot be over emphasized. It is analogous to the patient safety standard that requires that all specimen containers are labeled at the patient's side. Both of these practice standards are intended to assure that the specimen material in the container actually belongs to the patient identified on the container.

### Case with Error

A pathologists' assistant (PA) obtained breast biopsy containers and tissue cassettes on two patients and then placed both patients' materials on the gross dissection bench. One patient had a previous history of breast cancer and the second patient had no history of breast cancer.

The pathology results produced a result of cancer for the patient with no cancer history and a normal result for the patient with a previous history of cancer. Both patients were under the care of the same surgeon, who contacted the laboratory and questioned these results since these conflicted with the patients' differing medical histories and symptoms.

### Explanation and Consequences

The surgeon notified both patients and explained that further testing was necessary. This information only increased the anxiety for both patients.

The laboratory determined that it was necessary to conduct DNA testing in an effort to verify the patient identity of the tissue. It was necessary to perform these molecular genetic tests on four different specimens: the tissue biopsies from both patients that were collected on the same day, the tissue from the original biopsy of the patient with the previous history of cancer, and a blood specimen obtained from the patient with no previous history of cancer. The DNA results demonstrated that the patient with a previous history of breast cancer did indeed have a recurrence. For the patient without a previous history, the DNA results from a blood specimen matched the DNA results from the biopsy, which were negative for cancer.

In this situation, the PA failed to comply with key steps in the written procedure. First, she placed materials from more than one patient on her workbench. Second, she failed to verify the patient identifiers on the tissue cassettes with the specimen container and the requisition. Surgical biopsies are irreplaceable specimens, and results determine whether the patient has cancer or is at risk for cancer. It is absolutely imperative for staff to comply with defined procedures to assure the fidelity of the specimen with the patient at all times.

# CONFIRM PATIENT IDENTITY WHEN PROVIDING VERBAL RESULTS

In the vast majority of patient care situations, information technology enables the clinician to have ready electronic access to test results. However, there are circumstances where a patient is in critical condition and the provider does not have immediate electronic access to results. The clinician will then call the laboratory and request a verbal report of the results. At any given time, the laboratory is performing tests on dozens if not hundreds of patients. Patients with the same or very similar names can be undergoing testing in the same time frame. It is still necessary to obtain patient identifiers to assure that the correct results will be provided on the correct patient.

### Case with Error Averted

A patient, Henry James, was admitted in critical condition to the emergency department (ED). As his condition deteriorated, the ED nurse called the laboratory and requested verbal results on STAT tests but provided only the patient name. The technologist located several patients with that name and two of them were active cases on that day. She requested that the ED nurse must provide one other unique identifier, preferably the patient's medical record number, so that the technologist could access the correct patient record.

### Explanation and Consequences

Staff from patient care units are not aware of the large number of patients that the laboratory serves in any given day from many sites, inpatient, outpatient, and "nonpatient" (patient specimens received from off-site locations). The laboratory professional must be prepared to prompt colleagues and ask them to provide the required patient identification. This practice will minimize the risk that the incorrect patient's test results are accessed.

# STANDARDS OF PERFORMANCE

▓ The laboratory service is unlike any other health care service in that testing is performed in the absence of the patient. Thus, it is absolutely imperative to define procedures to assure that both patients and patients' specimens are accurately identified throughout the testing process. Staff must consistently comply with key performance standards for PPID.

▓ At the time of collection, a health care provider must perform both active and passive identification. In those circumstances where the patient cannot communicate, the laboratory should define appropriate actions in consultation with risk management. All labels must be applied to specimen containers in the presence of the patient.

▓ During the analytic phase, it is particularly important to design manual work processes to minimize the risk of erroneously placing patient specimens in an incorrectly labeled test container.

▓ All verbal communication must include unique patient identifiers. This will reduce the risk of confusing information about two different patients with the same or similar names.

# 3 Quality Management and Performance Improvement

## OVERVIEW

A defined quality management (QM) program is an essential tool to measure the success of clinical testing services. A QM program must evaluate the production of test results across the entire performance spectrum: the preanalytic stage, the analytic stage, and the postanalytic stage.

The clinical leadership is responsible for identifying specific performance indicators to monitor against defined performance thresholds. Generally, tests are selected for monitoring based on the potential impact to patient care. "High volume" tests are those for which errors would affect a large number of patients. "High risk" tests are those for which errors would produce serious negative outcomes, including loss of life or limb. Performance indicators should include appropriate activities across the preanalytic, analytic, or postanalytic phases.

Performance thresholds may be defined in several ways. A common practice is to base a threshold on the required outcomes for patient care, such as a turnaround time (TAT) of 40 minutes for test results ordered on stroke patients when timely therapeutic drug intervention is required. Some performance benchmarks are determined by the accreditation standards, for example, the presence of two unique patient identifiers on specimen containers.

There are some tests that largely serve an outpatient population. In this setting, it is necessary to understand the industry standards established in the commercial marketplace such as acceptable wait times for patients in an outpatient phlebotomy service.

## COLLECT AND ANALYZE DATA TO SUPPORT PATIENT OUTCOMES

To effectively evaluate routine performance of a test at any phase, one must first measure the activity and then define the performance standard. It is often necessary to conduct several observations and measure the time to complete the procedure from start to finish. Once the performance standard is determined, then the performance threshold can be defined. Generally, an acceptable performance threshold is reported as the ability of the laboratory to complete the procedure and meet the performance standard with a high rate of success. A poorly defined performance threshold or inadequate data collection will lead to a failure to identify problems with the procedure, and thereby pose an undue risk to patient care.

### Case with Error

The laboratory and neurosurgery staff were collaborating on the maximum TAT for STAT test results to determine whether stroke patients were eligible candidates for thrombolytic therapy. It was agreed that the maximum TAT for results would be set at 40 minutes. The TAT for tests in the stroke panel was added as a performance indicator to the monthly QM report. Data were collected and reported as an average, with the average actual performance at 38 minutes during the first quarter. Although the laboratory reported that its TAT performance was meeting the threshold, the clinicians objected based on their first-hand experience with a number of patient cases where results were provided beyond the 40-minute performance standard.

### Explanation and Consequences

The laboratory staff failed to properly analyze the data. Collecting the TAT for each test and then calculating an average meant that half of all tests must be above the average. In this case, the average TAT was 38 minutes compared with a maximum threshold of 40 minutes. Thus, the clinicians' anecdotal observations were valid since many of the test results exceeded the calculated average of a 38-minute threshold.

The intent of monitoring actual performance relative to a standard is to assure that the production process meets the standard. For this situation, it would be better to measure the number of instances where the TAT was met. Upon reevaluating the data in this manner, the laboratory would obviously see that its performance was not meeting the standard in approximately 50% of all cases. This analysis demonstrates that the laboratory leadership must take action to improve performance.

### Case with Error

The standard for providing a result on intraoperative surgical pathology consults is that 90% of all single consult requests must be completed in 20 minutes from receipt in the surgical gross room to notification to the surgeon. Data were collected, reviewed monthly during the laboratory's QM meeting, and the performance apparently met the standard. However, surgeons repeatedly voiced dissatisfaction that the TAT was excessively long.

### Explanation and Consequences

A delay in providing results for intraoperative consults extends the time and, thus, the associated risks to which a patient is exposed when undergoing a surgical procedure.

This laboratory failed to adequately measure its performance. Data were collected for a sample of all procedures with intraoperative consults, specifically only intraoperative consults performed on Mondays. The hospital surgical suites were fully scheduled every day. However, the case mix on Mondays was such that it produced the lowest number of intraoperative consults test orders. Further investigation into this matter found that the laboratory could only measure the TAT for those intraoperative consults when both the receipt time and report time were documented, and only 40% of the consults were completely documented with both times.

There are certainly circumstances where it is reasonable to assess performance on a sample of the total test population. When doing so, it is imperative to fully understand the population and properly apply statistical sampling methods. For this particular circumstance, the decision to evaluate a sample of the total population on a day when the test volume is low was further aggravated by the fact that less than half of the selected sample was being evaluated.

## DEFINE PERFORMANCE STANDARDS

▶ When there is a failure to accurately define a performance standard, then it will not be possible to identify problems and take corrective action. Various performance indicators can be monitored such as the TAT for test results or patient wait times. A threshold must be appropriately defined for each performance indicator. The actual performance can then be compared to the threshold and evaluated as to whether the actual performance is acceptable or unacceptable.

## Case with Error

Patient surveys demonstrated consistently high levels of dissatisfaction for wait times in the outpatient phlebotomy service. Survey comments also documented that some patients would leave and have the tests performed elsewhere.

In response to the survey data, laboratory management realized that if they improved patient satisfaction, they might regain the revenue lost to outside facilities when patients left. Data collection revealed that the average patient wait time was 30 minutes. Several actions were taken to improve wait time and the subsequent data collected demonstrated a decrease in average wait time to 20 minutes. However, patient surveys continued to show the same high level of dissatisfaction with the service and that patients continued to leave.

## Explanation and Consequences

Laboratory management failed to properly define the performance threshold. A decision was made to simply reduce the current wait time by 10 minutes with the assumption that this improvement would be acceptable to the patients. To compound the problem further, the collected data were evaluated by calculating an average wait time.

For some performance standards, it is best to study those practices where the performance threshold is met. When patient satisfaction failed to improve after initial actions, the laboratory management team conducted a survey of commercial laboratory outpatient phlebotomy services. It found that patients were generally satisfied when waiting less than 10 minutes. When waiting was between 11 and 20 minutes, patient satisfaction deteriorated, and after 20 minutes, patients would leave. With this information, an appropriate performance threshold of meeting a wait time of 10 minutes or less for 85% of all outpatient phlebotomy encounters was established. Commensurate with improved patient satisfaction scores, the laboratory also demonstrated increased outpatient test volume and revenue.

### Case with Error

The hospital leadership embarked on an initiative to reassign inpatient phlebotomy to nurses and patient care assistants at the bedside. To measure staff compliance with performing positive patient identification, the laboratory monitored the number of mislabeled specimens. The error rate was reported as a decimal and the performance threshold was set at an acceptable rate of less than 2 per 10,000 errors per month.

This quality data were reviewed monthly by a joint laboratory–nursing committee. During the first quarter of the new work process, the actual performance was just below the threshold. Nursing leadership viewed their performance as acceptable. However, both the laboratory medical director and manager reported that the staff frequently contacted patient care units to recollect specimens. This in turn was triggering an increase in the number of complaints from physicians that test results were delayed.

### Explanation and Consequences

The laboratory leadership incorrectly selected the measurement to be monitored. From a nurse's perspective, the staff on a unit may manage several dozen patients a month.

Thus, 2 errors out of 10,000 in a month appear to be very small. However, the laboratory is performing services for hundreds of patients every day and was generating more than 10,000 results per week.

A decision was made to report the absolute number of errors on a monthly basis rather than a decimal. With this change, the actual performance indicator became the number of mislabeled specimens. The new report format demonstrated that between 50 and 55 labeling errors per month were found to occur. Thus, almost two patients per day each month were subjected to an unnecessary second venipuncture. Nursing leadership then agreed that this performance was unacceptable. The committee embarked on a number of initiatives to reinforce staff compliance with positive patient identification procedures. It also successfully targeted a goal of reducing the number of errors by 50% and "holding the gain" for 6 months before embarking on a goal to reduce errors by another 50%.

# STANDARDS OF PERFORMANCE

■ Laboratory performance must be measured to ensure that the service meets or exceeds the defined standard. An appropriate performance threshold must be defined for the performance indicator.

■ Measuring performance indicators requires that a representative sample must be collected for data analysis. Therefore, it is necessary to understand the unique characteristics that are associated with a particular test or procedure.

■ It is important to consider the most applicable unit of measure when reporting a performance indicator. Many accrediting agencies frequently define acceptable performance as the ability to meet the performance standard at least 80% of the time.

■ Generally, reporting an absolute number for performance indicators is an appropriate measure to use where errors can have significant consequences such as an incorrect blood transfusion.

# 4 Financial Management

## OVERVIEW

Provision of laboratory services requires resources: staff, equipment, and supplies. The leadership is responsible to acquire resources and to manage expenses in both the operating and capital budgets. It is also necessary to manage the revenue stream so that billing claims are submitted correctly and timely to maximize payment for services.

When assessing the expansion of an existing laboratory service or implementation of new test programs, the financial impact of the operating costs (if necessary, capital expenses as well) must be calculated. This financial analysis should also consider other factors. The opportunity cost of not choosing to pursue new or expanded programs must be assessed. This evaluation should include the impact on patient care such as a decreased length of stay (LOS) by decreasing result TAT. A financial assessment should also consider the pros and cons of "make versus buy" and determine whether it is cheaper and more efficient to "make" the test in the laboratory or "buy" it from a vendor.

Operating expenses should be routinely monitored on a monthly basis. Supervisory staff should review budget reports for actual expenses and confirm that both staff salary and supply costs are correct. It is important to engage staff so that they can contribute to controlling supply expense. Employees should understand that payment for services is not provided until 45 days or more from the date when the claim is submitted. Defined processes for inventory management should be structured to minimize unnecessary overstocking on supplies.

Conversely, billable test volume and revenues must be regularly reviewed as well. Management oversight should include verification that the correct Current Procedural Terminology codes are submitted on claims. The timeliness of claim submissions should be monitored as well to minimize the risk of nonpayment for services rendered.

## MONITOR SUPPLY EXPENSES

▶ Supply inventory should be managed so that reasonable quantities are maintained on site. Excessive supply inventory ties up financial resources that could be better spent for salaries, capital equipment purchase, new or expanded programs, and infrastructure needs.

### Case with Error

A laboratory supervisor at a 400+-bed hospital, with a 40-bed pediatric unit, was offered a significantly discounted price for microtainer blood tubes. The vendor required a minimum order volume in order to meet the discounted price, so the supervisor placed an order for several thousand microtainers. The supply lasted for several years.

### Explanation and Consequences

The laboratory supervisor failed to understand that the payment to the vendor for these supplies would occur now but the laboratory would not receive payment for using these supplies until the tests were performed and the claims were submitted to the insurer. It would take several years to use these supplies. Also he failed to recognize that the vendor was motivated to meet quota and earn a bonus. It is imperative to effectively manage supplies so that they are purchased and paid for in quantities commensurate with actual use and in a time period that is relatively close to when the revenue payment would be received.

▶ The management team must routinely review actual expenses that are charged to the operating budget. Although much of the purchasing and accounts payable functions are electronically processed, there are still opportunities for error.

### Case with Error

A laboratory had contracted for several instruments through operating leases. A manager failed to review each expense line item and only reviewed the total actual expenses relative to total budget. Over a 6-month period, one of the operating leases was not paid and the vendor contacted accounts payable with the demand for full payment on the amount due plus penalty fees of 25%, as per the contract.

### Explanation and Consequences

The manager failed to see that the actual payment for operating leases was under budget because she did not review each expense line. Numerous factors can impact the operating budget, so it is necessary to review each actual expenditure and investigate the cause for either a positive or negative variance. It's important to understand the causes that contribute to expense variances so that any unnecessary penalties are avoided and informed decisions can be made for effectively managing resources.

### Case with Error Averted

The laboratory manager reviewed the monthly expense budget and identified a significant over expenditure for specimen collection containers. Actual test volume was at the targeted budget volume, so increased test activity did not support this increased supply expense.

These specimen collection containers were provided by the laboratory to the physician offices. The laboratory maintained the product inventory and would ship it upon request to the physician offices. Upon further investigation, the laboratory manager identified that several large multiphysician practices were requesting more supplies than specimens received for testing. The manager contacted the supply vendor who confirmed that her laboratory had increased its order for supplies, but that several other laboratories in the territory had decreased orders for the same supplies.

Given this information, the manager constructively engaged the office managers for these physician practices. These discussions revealed that the physician office staff appropriately collected specimens and sent them to the contracted laboratory as determined by the patient's insurance. However, the physician office employees were only requesting supplies from one laboratory. The laboratory manager and office managers agreed that the office staff were responsible for contacting each laboratory facility to replenish supplies based on the volume of specimen containers sent to each laboratory.

### Explanation and Consequences

The laboratory manager was diligent in reviewing both revenue and volumes and the associated expenses in the operating budget. This allowed her to identify a discrepancy between test volumes and revenues as compared with actual expenses. She avoided a potentially significant impact to operating expenses if this matter had gone undetected for a long period of time.

## MANAGE REVENUE

▶ In those few situations when a new test method is released with a new CPT code, it is important to verify that insurers will reimburse payment for the new test. However, there are circumstances when an existing test with a CPT code will be recommended as part of a new protocol for patient care diagnosis or treatment. This situation also warrants a review to determine that insurers will reimburse payment for the new use of the test.

### Case with Error Averted

The clinical laboratory was ready to implement HIV testing in conjunction with the Pap smear. This new testing program would support recently recommended preventive screening guidelines for cervical cancer. However, the laboratory manager recognized that the insurers may not be aware of the new guidelines and may not pay an insurance claim for both tests. To address this concern, laboratory leadership collaborated with the clinical chief of staff and the chief financial officer. A team was assembled to submit the necessary documentation for approval of payment for both tests.

### Explanation and Consequences

The laboratory management team took a proactive stance on this matter and requested direction and support from other departments. This action was to assure that the laboratory would receive payment to cover its operating costs. More importantly, the laboratory avoided a significant problem with its clients, both physicians and patients. Had the insurance companies denied payment on the claims, the hospital would have then billed the patients. This would have been a source of significant dissatisfaction for both the physicians and their patients.

> Managers should use financial reporting tools to monitor the revenue cycle. A commonly used report is one that monitors the number of denied claims. This report can indicate problems that require the manager to investigate and take action to assure that payments are received. Claims may be denied due to delays in submitting claims, incorrect CPT codes or modifiers, missing or incorrect diagnosis, and many other causes.

### Case with Error

A new manager recently joined a community hospital and began reviewing the financial summary report for denied claims. The laboratory was incurring a six-figure revenue loss for claims submitted on tests sent to its reference laboratory. In discussions with the finance staff, she was notified that these claims were denied payment because the bills were submitted too late.

### Explanation and Consequences

Contracts with insurers often include defined time periods for providers to submit claims and for the insurers to provide payment. Failure by either party to meet performance standards can trigger a penalty as defined in the contract, which may include nonpayment of claims submitted beyond a deadline.

The new manager investigated the existing billing process for tests sent out to reference laboratories. She found that over several years, a number of new reference laboratory tests had been added to the test order menu. However, the staff were using a manual process for billing these tests and also failed to provide the required data to enable automatic electronic billing. The volume of manually billed tests had increased to the point where the staff were frequently unable to complete this manual process before the specified deadline. To resolve this problem, the manager pulled together an interdepartmental team of both laboratory and finance staff to build and implement an automated billing process for these tests.

## ASSESS PROGRAM OPPORTUNITIES

▶ When evaluating the financial impact of a test, the potential effect on patient care must be considered as well. Upon initial examination, a test may have a modest financial impact on the laboratory budget. However, the results may allow providers to initiate treatment so that it substantially benefits both patient care and the institution's financial picture.

## Case with Error

A financial evaluation of a test was conducted. This test supported a largely inpatient population, whose insurance coverage provided a fixed, or capitated, payment. This test's impact to the bottom line would be to increase costs. Given the relatively high unit cost of performing the test and the current volume, a decision was made to perform the test three times a week. However, this test result was a key component in determining discharge decisions. Thus, the decision to "batch" the tests increased the LOS for inpatients.

## Explanation and Consequences

The laboratory manager wanted to achieve the lowest cost per test to realize the maximum cost savings. However, by "batching" this test, many patients were waiting 1 or 2 days for discharge while the test orders were pending until the next scheduled "run." A reevaluation of the financial impact demonstrated that the cost for performing the test on a daily basis was far outweighed by the lower overall discharge costs that were generated by decreasing the patient's LOS.

# STANDARDS OF PERFORMANCE

▦ Managers should routinely monitor supply expense and inventory. In addition, all laboratory staff must participate in effectively managing supplies to better control operating expenses. For every dollar in operating cash spent on supplies, one dollar less is available for salaries and capital equipment.

▦ The need to obtain the best price for supplies should be balanced with maintaining a reasonable inventory. Generally, payment for testing services is not received until 45 days or longer after the service has been performed. Thus, there is little benefit to investing operating cash to purchase a product that could last for many months or years.

▦ Operating budget reports should be reviewed monthly to assure that appropriate supply expenses are documented. This includes verifying that all lease payments or reagent rental fees are correct, as per the contracts.

▦ Product inventory should be monitored and this includes those test supply items that are provided to clients. For those supplies that are provided to clients, there should be a reasonable association between the test volumes returned to the laboratory with supply items delivered to the client.

▦ Operational processes should support accurate and timely billing and claims submission to assure that the full revenue payment is received for services rendered. When implementing new test programs it may be necessary to confirm that insurers will provide reimbursement for a new service.

▦ Automated electronic billing must be used as new tests are added to the test order menu. This will assure that claims are consistently and correctly submitted within contractual requirements.

▓ Financial management reports should be monitored to identify any problems with claims submissions and payment denials. The manager is responsible for taking corrective action to assure that claims are accurate, complete, and meet contractual deadlines.

▓ The laboratory leadership is responsible for recognizing the larger impact of the laboratory service on patient care. There may be circumstances when incurring additional costs to provide a laboratory service is more than offset by an enhanced outcome for patient care.

# 5 Staff Management

## OVERVIEW

The laboratory leadership is responsible for assuring that qualified staff are hired and properly trained to provide accurate test results. Building a team of competent employees starts with the proper selection of candidates.

A well-defined training program is necessary to assure consistent performance by all new employees. Orientation training must also include competency assessments to objectively measure actual task performance relative to the procedure standard.

Management must also define and communicate objective performance standards to staff. Employees should receive periodic reports so that they can clearly understand how they are completing their work as compared with the standard.

# SELECT QUALIFIED CANDIDATES

▶ The selection process starts with a well-defined job description that includes required education, experience, and skills. It is necessary to interview candidates and thoroughly assess their capabilities to meet the job requirements. However, it is equally important to evaluate the candidate's interpersonal communication skills with other employees and to obtain objective references from the candidate's current and previous employers.

### Case with Error

A pathologists' assistant (PA) was a candidate for a position with an active hospital practice. He was well qualified with work experience at both academic medical centers and research institutions. His letters of reference were provided from previous employers and confirmed that he demonstrated strong clinical skills. However, the hiring manager did not contact the candidate's current employer. The candidate stated that he did not want to jeopardize his position with the current employer if he was not selected by the hiring manager.

This candidate was hired and successfully completed a 90-day orientation. Subsequently, he was frequently absent and, after appropriate documentation and disciplinary action, was eventually fired less than a year after he was hired.

### Explanation and Consequences

By failing to obtain a reference from the current employer, the hiring manager was unaware that the PA had a poor attendance record with his current employer. The hiring manager then incurred the added expense of recruitment and training for a second time in less than 1 year. While the position was vacant, current employees then had the added burden of completing a larger caseload.

The hiring manager had offered the position contingent upon the successful completion of a background check. This check verified the candidate's education, certification credentials, and lack of criminal record. The manager could have offered the position with a second contingency: an acceptable reference from the candidate's current employer.

### Case with Error Averted

A candidate from a well-regarded out-of-state institution was interviewed for a laboratory manager position with a number of significant clinical services. The applicant's interviews with both the medical director and the administrator to whom the position reported went well. This applicant was also interviewed by other individuals, both peers and subordinates, who would routinely interact with the person in this management position. These interviews were markedly different. The candidate demonstrated poor interpersonal skills such as dominating the interview meeting with subordinates and using unprofessional language.

### Explanation and Consequences

This institution conducted its interview process to engage staff at all levels (superiors, peers, and subordinates) to evaluate candidates. In this instance, the candidate's interview performance most likely reflected actual work performance: he effectively "managed" his interactions with superiors, but did not constructively engage coworkers and subordinates. Using this "group" interview process enabled this laboratory team to better assess the candidate's interpersonal and communication skills.

# DEFINE STANDARDS AND MEASURE ACTUAL PERFORMANCE

▶ Performance standards should be clearly defined and communicated to staff so that they understand what is required to successfully accomplish tasks. It also enables coworkers to form a stronger sense of team since their work is measured based on objective standards and not subjective perception.

### Case with Error

The specimen receiving department defined a standard to process specimens in 10 minutes from time of receipt to delivery at the workbench. Actual performance data indicated that the staff were consistently failing to meet this turnaround time standard. This performance failure impacted the ability of the laboratory to meet the turnaround time requirements for STAT test results. In turn, management was faced with addressing numerous adverse event reports from critical care units and the emergency department regarding delays in test results.

### Explanation and Consequences

Management defined an acceptable performance standard for its operation, but employees were not accountable for meeting this standard. To address this problem, the manager requested a report from the laboratory information system (LIS) that documented the work performed by each employee, based on his or her computer log-on. Each employee received regular updates on his or her performance. In addition, the manager "blinded" each employee's identity and then posted a report with every individual's performance. Each individual knew his or her own performance identifier and was able to compare his or her individual performance to both the standard and coworkers' performance. Within 6 months, all employees were performing at a level that either met or exceeded the minimum standard. Consequently, the laboratory improved its turnaround time for STAT test results.

### Case with Error Averted

The specimen preparation laboratory in cytology had a defined orientation and training program for newly hired laboratory assistants. This job required a high school diploma and a minimum of 1 year's previous laboratory experience. A laboratory assistant in the clinical laboratory applied for a transfer to the cytology specimen preparation laboratory and was accepted. This employee had 5 years of experience in the clinical laboratory and a demonstrated record of above-average performance. After only 1 month of the 3-month orientation program, the employee was unable to meet the training program requirements. The cytology specimen preparation requirements were more complicated than the specimen preparation tasks that she performed in the clinical laboratory. She requested a transfer to her previous position in the clinical laboratory and this request was accommodated.

### Explanation and Consequences

The cytology training program provided each trainee with a checklist of tasks and required competencies. This tool clearly communicated objective criteria. It enabled the trainee to measure her performance and assure that every trainer addressed all required tasks.

A structured training program allows the new employee to better understand the performance standards that must be met. In this particular situation, the employee could recognize that her skills were not well matched to the new position. She was able to return to her previous position and avoid the negative outcome.

## ADDRESS COMPENSATION ISSUES

▶ Many factors can impact the ability to hire and retain staff. A competitive salary is one of the key elements for attracting and retaining employees. Laboratory management must identify those circumstances that can create a noncompetitive position with salaries. It is also important to engage the human resources staff to assist with collecting data and, if necessary, finding solutions.

### Case with Error

A laboratory at an academic medical center was unable to recruit a master's degree, board-certified PA. Qualified candidates routinely rejected job offers because the salary was unacceptable. When the medical director and manager presented this information to the employment recruiter, they were informed that the salary range for the PA position was competitive in the regional marketplace.

### Explanation with Consequences

The human resources market data were based on salary ranges in the region. However, the regional hospitals did not employ staff with the equivalent training and certification to that desired by the academic institution. The laboratory leadership met with the human resources director and explained that the community hospitals were employing individuals with either an associate's or bachelor's degree who were then trained on the job. At the medical center, there was a much higher acuity of patient case mix, particularly because it supported a federally designated comprehensive cancer center. This supported the need for highly trained PA staff.

Both the laboratory and the human resources staff collaborated to collect salary data in the national marketplace. With this information, salary ranges were adjusted appropriately to compete in the national marketplace, and qualified candidates were hired.

### Case with Error

A large academic medical center in a metropolitan area was experiencing a high vacancy rate for its medical technologist positions. This vacancy rate had almost doubled from 8% to 15% and was reached in a relatively short time of 4 months. A number of employees had resigned and accepted employment at other competing institutions.

### Explanation and Consequences

As would be expected, this shortage of qualified staff in the market subsequently produced demand for higher salaries to induce qualified candidates to leave their current employers.

In this specific situation, the laboratory management worked with the human resources staff. Exit interviews were conducted with those employees who submitted resignations to document their reason for leaving: better salary. The human resources staff also collected salary data for the market. A new salary structure was defined, approved, and implemented. With the new salary structure and an aggressive recruitment campaign, laboratory management was able to retain its current employees and successfully recruit applicants for the vacant positions.

# ASSESS STAFFING LEVELS AND WORKLOAD

> ▶ It is important to monitor test volume activity and identify when new staff are required. This allows the laboratory to continue to meet performance standards and minimize any disruption in service.

### Case with Error Averted

The laboratory manager routinely monitored test volume, and increased activity indicated the need for an additional technologist position. This laboratory operated 24/7, and volume data were measured on an annual basis. Since the majority of the test volume occurred on the day shift, it was initially assumed that the position should be added on the day shift. The manager collected test volume activity per shift for each day of the week from the base year and compared it with the increased activity in the current year. This analysis demonstrated that the largest volume increase occurred between 4:00 AM and 12:00 PM on Tuesday through Saturday.

### Explanation and Consequences

Several new programs in different clinical disciplines had been added at this hospital. The increased test volume was driven by the practice patterns of the physicians. By conducting a careful analysis, the laboratory manager was able to add staffing resources when they were most needed.

# STANDARDS OF PERFORMANCE

▓ Evaluating candidates should include an assessment of their clinical expertise and their communication and interpersonal skills. When possible, candidates should be interviewed by colleagues and subordinates as well as by superiors.

▓ It is recommended that references are obtained from both current and previous employers when an applicant is the preferred candidate for a position.

▓ Clearly define objective and quantifiable performance measures for employees. This allows employees to understand what is expected and enables the laboratory to dependably support patient care.

▓ Identify conditions in the marketplace that can affect salaries and impact an employer's competitive position. The human resources staff can provide the necessary data to justify appropriate actions to recruit and retain staff.

▓ Monitor test activity to determine staffing needs. When appropriate, analyze data and define when staffing coverage should be assigned.

# 6   Laboratory Safety

## OVERVIEW

The laboratory leadership team must assure that the work environment complies with safety standards as defined by government regulations, accreditation standards, and institutional policies. In addition to implementing procedures, the staff must be educated to properly perform safety procedures and comply with them.

## DEFINE PROCEDURES AND MONITOR COMPLIANCE

Since the mid-1980s, the Occupational Safety and Health Administration (OSHA) has required that personal protective equipment must be used in all situations where there is a risk of biohazard exposure. In the laboratory, all staff are required to wear laboratory coats and gloves when handling specimens and performing tests.

### Case with Error

A small community hospital laboratory was undergoing an on-site inspection for reaccreditation to maintain its CLIA license. The inspectors observed that a number of employees in several laboratories failed to wear laboratory coats. A component of the accreditation standards included the requirement to wear laboratory coats for protection from biohazardous contamination. The inspecting team cited the laboratory for failure to comply with this standard.

### Explanation and Consequences

The core laboratory service was located in the original rooms that had been built more than 40 years earlier. As testing technology changed over the decades and new instruments were installed into this space, the air conditioning system was not upgraded to properly maintain an appropriate ambient temperature. Although the inspectors concurred that the environment was uncomfortably warm, this circumstance did not justify an exception to the safety regulations.

This failure to manage the environment placed both the employees and the institution at risk. Clearly, the employees were at daily risk for personal exposure and contamination of both the work environment outside the laboratory as well as their home environment. The institution was at risk for significant fines that could be levied by both federal and state agencies.

In this situation, it was incumbent upon the inspectors to cite this deficiency and for the accreditation agency to hold the employer accountable for compliance.

> ▶ Laboratory management must have defined written procedures for laboratory safety and must monitor compliance with these procedures.

### Case with Error

A laboratory had a well-written procedure for the appropriate storage of hazardous chemicals. This procedure had been in place for a number of years and had been reviewed and updated as necessary. A new laboratory supervisor was hired. Shortly after her arrival, the new supervisor conducted a review of the chemicals in storage and identified a highly dangerous situation. Picric acid had been stored beyond an acceptable limit and had become crystalline.

### Explanation and Consequences

This laboratory's procedure did require routine review of hazardous chemical storage and documentation of such review. It was common practice in this facility for the staff within each laboratory section to conduct a self-review for safety compliance. A more effective practice is to establish a formal review that can be conducted by a team comprising staff from relevant departments within the institution such as infection control and environmental safety. When conducting an inspection, the team should use a written checklist to document compliance with all applicable safety requirements.

> ▶ Staff are required to comply with all safety procedures, and laboratory leadership is responsible for holding staff accountable to properly perform work in accordance with the safety standards.

### Case with Error

A new federal regulation requires that manufacturers must modify the method for replacing blades in surgical scalpels to reduce the number of laceration injuries. A laboratory evaluated vendors' products and selected a device to remove dull blades and install a new blade onto the scalpel. A pathologist preferred not to use the device and continued to manually replace surgical blades onto the scalpel. She suffered a significant laceration that required stitches and the injury was reported to Worker's Compensation.

### Explanation and Consequences

The physician's actions placed both herself and her employer at unnecessary risk. Safety regulations, whether defined by government agencies or institutional policies, are applicable to all employees, regardless of status.

In this situation, the staff observed the physician's noncompliant behavior, but were uncomfortable with reporting this safety breach. The laboratory leadership engaged in an active effort to educate all staff about the importance of reporting noncompliant performance to appropriate laboratory or institutional departments. The staff were informed that they were protected from any retaliatory actions for reporting these problems. In addition, there was an initiative to educate all staff that compliance with safety standards was mandatory and not optional, and that failure to comply would result in appropriate disciplinary action.

## STANDARDS OF PERFORMANCE

▨ Laboratory management is obligated to implement defined procedures that meet safety requirements and monitor staff compliance with safety standards. Appropriate staff resources within the laboratory and from outside departments should be actively engaged with defining, implementing, and monitoring laboratory safety matters.

▨ Safety practices are applicable to all employees who perform those tasks that are covered by government regulations or institutional policies. As new safety initiatives are implemented, staff should be trained in proper practice and understand that there are consequences for noncompliance. It is management's responsibility to monitor compliance and provide a mechanism to allow staff to report safety failures without fear of retaliation.

# Specimen Logistics

## OVERVIEW

An important step in the preanalytic process is the transport and delivery of specimens to the laboratory bench. It is imperative to understand the regulations that govern specimen transport and assure that the actual practice is in compliance. The transport process must be designed to minimize the risk of losing specimens as they are moved from the collection site to the laboratory bench.

There are Department of Transportation regulations that define numerous standards for transporting biohazardous materials. These regulations apply to both internal transport within the hospital from the patient "bedside" to the laboratory and external transport by courier services from an outside facility to the laboratory. A health care facility is accountable for contracted vendors. Therefore, it is incumbent upon the laboratory management to assure that its contract with an outside courier service requires vendor compliance with regulations. The laboratory will be liable should a contracted courier service fail to meet Department of Transportation standards.

There should also be an efficient process to assure that specimens are moved from the collection site to the laboratory bench. This process should be designed to move the specimens in a timely manner. Within a health care facility, it is generally easy to accomplish timely transport through the use of pneumatic tube transport systems. The widespread adaptation of computerized provider order entry enables the laboratory to better manage pending test orders with the receipt of patient specimens. However, this process design is more challenging when moving specimens from satellite collection facilities to the laboratory.

## TRANSPORT SPECIMENS FROM SATELLITE SITES

▶ Specimens must be transported, at all times, in a manner that complies with all safety standards and minimizes risk of exposure to biohazardous materials.

### Case with Error

A laboratory had expanded the number of satellite collection sites in response to a recent contract to support a large multi-specialty physician practice with a dozen locations. The laboratory's contracted courier service was notified that additional routes and deliveries would be required. Shortly after implementing the service to the new clients, laboratory staff reported to the manager that the courier staff were delivering specimens in "large plastic trash bags." The manager then contacted the courier service and reported this noncompliant activity and requested that they correct it.

### Explanation and Consequences

Federal regulations require that all specimens are transported in sealed containers that must be properly labeled as containing biohazardous material, maintaining proper temperature during transport to assure specimen integrity. When expanding its service, some of the courier staff did not comply with the company's defined procedures for transporting biohazardous materials. This failure placed both the vendor and the laboratory at risk for liability if there had been a leak of any biohazardous material.

> A process for "handing off" specimens from one location to another must be defined for both routine and nonroutine circumstances. Any number of communication tools (verbal, written, or electronic) can be appropriately applied to the situation in an effort to minimize the risk of losing specimens.

### Case with Error

A body fluid specimen was collected in a satellite outpatient facility with a STAT order for laboratory testing. Laboratory staff documented receipt of the specimen on the log sheet and notified the contracted courier vendor that a STAT specimen transport was necessary. The courier vendor picked up the STAT specimen and delivered it to the laboratory. He placed it in the designated "specimen delivery bin" and departed. Several hours later, the ordering provider contacted the laboratory for results, at which time the laboratory identified that the test had never been ordered and the specimen could not be located.

### Explanation and Consequences

Failure to perform this test clearly impacted the patient in two ways. The diagnosis and treatment decision were delayed, and the patient was inconvenienced as he had to return to the outpatient facility.

The laboratory practice failed to define a process for the infrequent occurrence of a nonroutine delivery outside of the regularly scheduled courier delivery times. Since it was an uncommon practice to have a STAT test with delivery outside the scheduled times, a change in practice was made. Rather than depositing the specimen in the laboratory's "specimen delivery" bin, the contracted courier staff must physically present the STAT specimens to the laboratory personnel.

# DEFINE EFFICIENT WORKFLOW PROCESS

▶ Workflow should be designed so that staff can complete tasks with consistent efficiency. When possible, nonvalue-added tasks should be removed from the work process.

### Case with Error

A specimen from a pediatric patient was delivered to the laboratory with test orders for molecular infectious disease testing. It was received on the evening shift of a Saturday, and the test would not be performed until Monday. The clinician contacted the laboratory for results on Sunday and was told that the laboratory would not perform the test until the following day, unless the clinician obtained approval from the laboratory's medical director. At the time when the necessary approval was obtained, the laboratory staff were unable to locate the specimen. The patient was subjected to another venipuncture in order to complete the necessary testing.

### Explanation and Consequences

An investigation subsequently located the original specimen, but this occurred after the second specimen had been obtained and tested. This laboratory occupied a large physical space, and the existing process required staff in central receiving to correctly deliver specimens to any one of 44 locations across the laboratory. In this case, the pediatric specimen had been delivered to the wrong laboratory location. However, given the large number of locations and physical space across which these specimen "drop-off" sites were distributed, it was difficult to quickly locate the missing specimen.

The management team redesigned its workspace and process so that two locations would each serve as a hub for specimen delivery. Specific subspecialty laboratories were assigned to each hub, and both room temperature and refrigerated storage were provided at each hub.

### Case with Error

A large laboratory was receiving increased complaints from critical care units that STAT test orders were not meeting the promised TAT of less than 1 hour. In addition, the laboratory's quality monitors also documented that there was a decrease in the number of STAT tests that were completed on time.

### Explanation and Consequences

This deterioration in the laboratory's performance for STAT tests was contributing to unnecessary delays in the diagnosis and treatment of patients. The management team collected data to evaluate the testing process from time of receipt to the time of result.

Upon analysis, it was apparent that there were delays at the initial step of specimen receipt. The same staff who performed all specimen receipt activity were also frequently interrupted by phone calls. While the laboratory certainly had to answer phone inquiries, this activity is a "nonvalue-added task" for timely specimen processing. A separate "workbench" was established and staffed with dedicated employees to answer the phones during peak periods. Staff could then focus on specimen processing without the distraction of answering the phones.

## STANDARDS OF PERFORMANCE

▦ All specimen transport activities must comply with government regulations to assure that there is minimal risk of biohazardous exposure or contamination. Laboratories are responsible to assure that both their employees and contracted vendors acceptably perform these duties.

▦ The procedure for transporting specimens should be designed to assure that there is documentation of a specimen's location as it moves through the "transport system." Documentation can be either electronic or paper and when necessary, may need to include verbal communication as well.

▦ The workflow process should be efficiently designed to move the specimen from the preanalytic step to the bench for analysis. Nonvalue added tasks should be removed and reassigned to designated staff.

# 8  Test Utilization

## OVERVIEW

In the ongoing national debate concerning the growing consumption of health care services, increased expenditures for laboratory tests, particularly molecular diagnostics, face continued scrutiny to control costs. Evaluating test utilization should cover both inpatient and outpatient activity as well as those tests that are performed in the laboratory and those sent to an outside vendor.

Effectively managing test utilization produces both operational and financial benefits. By reducing excessive orders for tests, there is available capacity, with both staff and instruments, to implement new tests as the need arises and to absorb increased volume. It will also increase the net revenue margin for test services that are reimbursed through capitated payments.

## CREATE PARTNERSHIPS WITH KEY CLIENTS

▶ In its effort to appropriately manage test orders, the laboratory leadership should consider engaging external resources. These resources can include practicing physicians who are "thought leaders" within a medical specialty, information technology tools that can monitor activity, or consultative expertise from other areas such as finance, compliance, legal, or risk management.

### Case with Error

The laboratory leadership conducted a review of tests sent to reference laboratories. These data indicated that more than 40% of these orders were placed during inpatient stays. However, the results would not be available until postdischarge. Since these tests were ordered during inpatient stay, the cost for them was included in the capitated payment for the admission.

### Explanation and Consequences

Although the results were valuable in the ongoing care of the patient, they were not necessary for inpatient diagnostic and treatment needs. Per regulatory requirements, since these tests were ordered during inpatient stay, they were covered under the capitated payment. Ordering these tests during the inpatient admission directly contributed to increased costs for this admission and, thus, lowered the net revenue.

The laboratory leadership collaborated with the hospital's medical staff leadership and initiated an educational program to instruct providers on the appropriate utilization for esoteric testing. Also, the services of the information technology (IT) department were engaged to modify the computerized test order entry program to redirect these test orders to the outpatient setting.

### Case with Error

The microbiology laboratory at a tertiary academic medical center observed an accelerated growth in the number of blood culture tests. This increased test activity exceeded the expected volumes based on actual inpatient admissions. Analysis of the data indicated that orders were placed, but the clinical symptoms did not support the need for blood culture and that duplicate orders were made by providers from different specialties.

### Explanation and Consequences

The excessive ordering of blood cultures created several problems. It unnecessarily subjected the patient to additional venipuncture and loss of blood. Laboratory resources, both staff and supplies, were inefficiently engaged in performing duplicate testing that had no value in patient care. The consumption of these additional resources increased the hospital's cost for these inpatient admissions and, thus, lowered its net revenue margin.

A multidisciplinary team was established with members from the laboratory, the infection disease division, and software programmers. The clinical experts developed a quick reference "pocket guide" card with an algorithm defining the appropriate clinical indications for ordering blood cultures. It was disseminated through an education program for the entire medical staff, with particular focus on the house staff. The IT staff modified the test order entry software program to generate a dialog box that notified the provider if there was already an existing order for blood cultures placed within the previous 24 hours. After these actions were implemented, an audit of blood culture tests was conducted and the data indicated reduced utilization.

# MANAGE TEST UTILIZATION

> ▶ Inappropriate utilization of some tests may occur because there is a poor understanding of the appropriate clinical indications. The laboratory clinical leaders can actively manage appropriate test utilization and constructively support the providers.

## Case with Error

An audit of coagulation test activity demonstrated increased volume for tests used in the diagnosis of both bleeding and thrombosis. This test volume was higher than expected for the patient case mix. Further analysis of the data indicated that the patients' symptoms did not support the clinical necessity to perform all of the ordered tests.

## Explanation and Consequences

Physicians were admitting patients with bleeding disorders of unknown etiology and ordering a number of esoteric coagulation tests. In an effort to promptly treat these patients, the physicians had developed a practice habit of ordering a broad spectrum of tests upon admission. They then sorted through this large number of results to make a diagnosis and initiate treatment.

This "shotgun" approach to coagulation test orders meant that many tests were performed unnecessarily, which produced some unintended consequences. These unnecessary tests contributed to higher costs. This practice also inappropriately utilized staff and instrument capacity that could have been available for performing necessary tests. The medical director collaborated with clinicians and jointly developed a program of algorithmic test selection to order only the correct and necessary tests.

## Case with Error

A hospital laboratory instituted the in-house performance of vitamin D testing. Previously, this test had been sent to a reference laboratory, but the volumes had increased such that it was now cost-effective to perform it in the laboratory. It was observed that there was an unexpected increase in requests for fractionated D2 and D3 levels.

## Explanation and Consequences

A more detailed analysis of the data revealed that some physician practices were routinely ordering the reference laboratory test to obtain the fractionated D2 and D3 vitamin levels instead of the vitamin D test offered by the hospital laboratory. The physicians had routinely ordered this more esoteric test rather than the routine vitamin D screening test, and did so even when the assay for total vitamin D was brought in-house. The laboratory medical director met with the physician leaders of the practices, reviewed the clinical indications for both tests, and test order patterns were appropriately modified.

# STANDARDS OF PERFORMANCE

▨ Patient care providers can obtain large amounts of diagnostic information from numerous tests. Whenever appropriate, the laboratory leadership should engage technology to assist in appropriately directing the selection and ordering of tests.

▨ Laboratory management must be effective stewards of its employees, instruments, and supplies. Test order patterns should be periodically monitored to identify any excessive or unnecessary utilization. Appropriate test order activity will assure that laboratory resources are efficiently supporting patient care and that there is available capacity to manage increased volume and implement new tests.

# 9 Competitive Performance in the Outreach Market

## OVERVIEW

Hospital laboratories have long recognized the opportunities of testing services in the outpatient market. The hospital has the capital infrastructure and capacity in the off-hour shifts that coincidentally is the time frame when most outreach testing is performed. A hospital laboratory also has an existing relationship with the physicians who admit patients and can build on that relationship.

Hospital laboratory leadership must recognize that performance standards in the outreach market are decidedly different from those for their inpatients. To succeed in outreach, the hospital laboratory must compete with commercial laboratories and at least meet, if not exceed, the performance standards in the outpatient marketplace. It is imperative to understand the service standards, conduct a SWOT (strengths, weaknesses, opportunities, and threats) analysis, and implement the necessary processes to meet the clients' service expectations.

# UNDERSTAND SERVICE REQUIREMENTS

▶ The laboratory management team often attempts to capitalize on its "strength" of clinical expertise. Frequently though, the laboratory fails to adequately evaluate the clients' service needs or provide the clinical consultation most needed by the ordering physician.

## Case with Error

When a national organization recommended a new program for cervical cancer screening, the laboratory developed the recommended screening services that included both the Pap smear and HPV tests for patients 30 years of age and older. However, these tests were performed in two different laboratories, cytology and molecular infectious diseases, that produced two separate test reports and the two reports were delivered on different days. The physician clients expressed dissatisfaction with this inefficiency of multiple result reports.

## Explanation and Consequences

The clients expected to read a single report containing both test results so that they can quickly determine if any further treatment is necessary. By providing two separate reports on different days, the ordering physician had to manually search for both reports to interpret results and make a treatment decision.

The laboratory failed to recognize this potential problem when it initiated its expanded test service. A number of process changes, including software modifications, were made to produce a single report with both results.

### Case with Error

A laboratory provided outreach testing services to several hundred physicians' offices. Many of these clients consisted of large physicians' practices with several office sites, and most of the physicians practiced at more than one site. The patients routinely selected to be seen at one site and thus, their medical records were stored at that one site. However, the laboratory delivered all patients' test result reports only to the primary office site. The physician office staff then found it necessary to sort reports and distribute them to the appropriate office where the patients were seen and their records were located. This process failure caused dissatisfaction with the large multisite physician practices.

### Explanation and Consequences

The laboratory failed to identify this important performance standard. In the early stages of its outreach initiative, the laboratory served physician practices at single sites. Its software was programmed to sort reports by ordering physician only. At first, this programming logic worked more effectively than using an address since different practices were located in the same professional building. To resolve this problem, the laboratory assigned a unique "client identification number" to each physician practice. The software could then sort test reports by client number for delivery to the correct location.

### Case with Error

In an effort to support a hospital's expanded women's health program in gynecology, the laboratory aggressively marketed its outreach program for cervical cancer screening and successfully captured approximately one-third of the market in the metropolitan service area. However, there were complaints from clients that the Pap smear result's TAT was not consistently maintained. The physicians stated that patients were notified that their results would be available in approximately 1 week. However, there were numerous occasions throughout the year when the laboratory results were reported from 10 to 14 days later than the standard TAT. Patients were dissatisfied when they contacted the physician's office and results were not available.

### Explanation and Consequences

The gynecologists were rightfully concerned that their patients inappropriately blamed them for the laboratory's performance problem. The laboratory manager collected data to confirm that the standard performance for result TAT in this market was 1 week. He also reviewed the result TAT for the previous year and identified a pattern. TAT exceeded 1 week whenever there was suboptimal staffing. The causes for staffing problems ranged from approved absences for medical leave or vacation time to vacant positions.

To resolve this problem, the manager calculated staffing coverage and included a vacancy factor. He created several part-time positions and developed a pool of part-time cytotechnologists who would be scheduled to work when necessary. After making these staffing changes, the manager regularly monitored the TAT and was able to consistently meet the performance standard.

## MANAGE STAFF PERFORMANCE TO SUPPORT OUTREACH MARKET

There are many service demands in the outreach market that do not coincide with service demands for inpatients and hospital unit staff. It is incumbent upon the laboratory leadership to understand the outreach service requirements, educate the staff, and implement any necessary workflow changes to support the clients' service needs.

## Case with Error

A hospital embarked on a program to establish a number of satellite facilities to support outpatient care. In addition to a broad array of specialty physician offices, various ancillary services were offered at these locations, including a phlebotomy service. A number of hospital-based phlebotomy staff transferred to these new phlebotomy positions at the satellite facilities.

During the first few months of operation, there were numerous complaints that patients were contacted to return for repeat venipuncture because the initial specimen was rejected due to insufficient quantity.

## Explanation and Consequences

The laboratory's phlebotomy staff failed to understand the impact of improperly collecting specimens. In the hospital setting, if a phlebotomist encountered a difficult venipuncture, she would endeavor to obtain as much specimen as possible. The specimen tubes were then delivered to the laboratory and a technologist would evaluate if there was an adequate quantity. If not, one of the designated "expert" phlebotomists would return to the inpatient unit and perform a repeat venipuncture to collect enough specimen for testing. This existing work process for the hospital setting was unacceptable in the outpatient setting.

The leadership team redesigned the work processes at the satellite sites. Staff were retrained to further strengthen their venipuncture skills. They were now required to immediately contact the "expert" phlebotomist whenever the first venipuncture produced unsatisfactory specimens. In addition, the patients' physicians agreed that they should be contacted on those rare occasions when even this "two phlebotomist" procedure failed to produce adequate specimens. The physician could then prioritize those tests that must be performed based on the available volume of specimen.

### Case with Error

During the off-hour shift, a key instrument malfunctioned and was down for several hours. When it became operational, the staff completed all of the tests ordered for inpatients and postponed the testing for the outreach program to the day shift. A number of the physician clients were scheduled to receive an automatically generated print run of result reports at their offices by 7:00 AM. Numerous outreach clients were dissatisfied with the failure to receive reports and contacted the laboratory.

### Explanation and Consequences

Laboratory management failed to define contingency plans when there was a disruption to operations. These contingency plans should address numerous unexpected circumstances such as instrument or LIS failures, courier problems with specimen transport, and staffing shortages.

The leadership developed an expanded training program and educated staff about the performance requirements for outreach clients. New protocols were implemented for notifying responsible on-call managers who could assist with resolving unexpected operational problems for outreach clients.

## DEFINE INFRASTRUCTURE REQUIREMENTS

▶ Hospital laboratories can engage support for the outreach market from a broad array of services provided by other departments such as finance and marketing. It is management's responsibility to constructively engage colleagues in other departments and clearly communicate service needs and to assist with development and implementation.

### Case with Error

The laboratory outreach program relied on staff from the finance department to complete patient registrations in the hospital's billing software. During the start-up phase, the finance department hired a registrar to work the evening shift and complete all registrations. As volume increased, the registrar was unable to complete all work by the end of the shift. When possible, the technologists would try to assist the registrar and, frequently, the registrar worked overtime. The laboratory performance was impacted because there were an unacceptable number of billing errors.

### Explanation and Consequences

The laboratory manager failed to properly monitor the billing process and maintain communication with her peer in finance. To resolve this problem, she collaborated with the finance manager and identified several action items. The laboratory manager developed a status report for the outreach program that notified support departments when new clients were added. She revised this report to include the estimated test volume that was expected from a client. The finance manager defined a performance measure and used it to monitor the volume of work and identify when additional staff were required. A quality control program was implemented by the finance department for outreach client billing to monitor the error rate.

### Case with Error

A hospital laboratory outreach program had established a small sales force. The sales staff designed promotional materials that were distributed to potential physician practice clients. A physician colleague notified the laboratory medical director that the promotional materials contained some inaccurate information about tests. The laboratory medical director immediately recognized that this situation could damage the laboratory's credibility with potential clients.

### Explanation and Consequences

The laboratory staff failed to recognize that they did not have the expertise to properly develop marketing tools such as brochures and pamphlets.

To address this matter, the laboratory leadership defined a procedure for the development, review, and approval of all promotional materials before these were released to the public. The laboratory manager engaged a colleague from the hospital's marketing department and requested assistance. A team to design promotional materials was formed with representatives from the laboratory's sales force and the marketing department. These materials were submitted to senior leadership in the laboratory and marketing for review and approval.

# STANDARDS OF PERFORMANCE

▓ Once the specific laboratory services for the outreach market have been identified, it is imperative to evaluate the service requirements as defined by the clients. Service requirements can be very broadly defined from ease of test ordering and availability of specimen collection supplies to complete and comprehensive reports that are consistently delivered in a timely manner.

▓ Management must assure that there are adequate resources, both instrument and staff, so that clients' service needs are reliably met.

▓ The management team must educate staff to understand that the service needs for the "outreach" market are different, and problems must be appropriately addressed. When there is a failure at any step in the process (preanalytic, analytic, or postanalytic phases), there should be a defined procedure that directs staff to notify the appropriate individuals who can assist with troubleshooting and problem resolution.

▓ Competing in the "outreach" market often requires infrastructure support outside of traditional hospital laboratory operations. The laboratory manager must enlist key departments, such as finance, IT, and marketing, and engage them so that they fully understand the performance requirements to meet the clients' service needs. Laboratory leadership should maintain constructive and ongoing communication with these external departments so that they can continue to support the service demands for the "outreach" clients.

# Selection and Management of Reference Laboratories

## OVERVIEW

The volume of tests sent to reference laboratories generally comprises a relatively small percentage of total ordered tests. However, the expenses incurred for reference laboratory services often constitute a significant portion of the supply budget. With the introduction of molecular diagnostic tests, the costs for reference laboratory testing have been growing exponentially for more than a decade. Laboratory leadership must be actively engaged in the selection of vendors for reference testing service and the providers' utilization of this resource.

# MONITOR UTILIZATION OF REFERENCE LABORATORIES

> ▶ For hospital-based laboratories, there are regulatory requirements which stipulate that the selection of reference laboratories must be determined jointly by the laboratory leadership and the hospital's medical staff. It is necessary for laboratory leadership to actively engage in the selection and use of reference laboratory facilities. The laboratory management team possesses the expertise to carefully evaluate the clinical quality, suitability of the test menu to meet patient needs, and service performance of the various vendors. These factors, in addition to cost, must be assessed when selecting a reference laboratory.

## Case with Error

A hospital laboratory generated more than 75,000 esoteric tests referred for off-site testing. More than 85% of the total volume was performed for outpatients. The laboratory was sending these tests to several dozen outside facilities. In less than 5 years, the laboratory expense for reference testing had grown by 75%.

## Explanation with Consequences

The laboratory did not control the selection of reference laboratories. Instead, the ordering physicians were selecting both the test and the reference facility. The allocation of tests across so many vendors prevented the laboratory from negotiating a discounted price based on volume.

An audit of these tests revealed two primary findings. First, various physicians were ordering the same tests, but selecting different reference laboratories. Second, the laboratory's primary vendor for reference testing offered an extensive test menu and could perform many of these tests.

To resolve this problem, the laboratory leadership defined a two-step process. The laboratory's clinical leadership obtained approval from the hospital's medical executive committee to pursue a program to consolidate all appropriate testing with the primary vendor. Then, the laboratory manager collaborated with the hospital's purchasing agent to negotiate a revised contract with its primary vendor and obtained reduced pricing based on increased volume. The laboratory easily achieved the volume targets and thus, was able to significantly reduce its operating expense for reference laboratories.

## EVALUATE VENDORS' FULL SCOPE OF PERFORMANCE

Given the explosive growth in molecular testing, managers are finding that costs for reference testing can consume 10% or more of total supply costs. It is certainly essential to consider costs when evaluating vendors for reference laboratory testing. However, clinical quality and service performance must also be considered. The vendor must meet defined clinical and service criteria so that the patient care needs are appropriately provided. Also, regulatory requirements hold the hospital laboratory accountable for any subcontractor's performance.

### Case with Error

A large community hospital laboratory proceeded with a competitive bid to several vendors for reference laboratory services. Most of this test volume was generated from the hospital's outpatient population. The hospital laboratory selected the vendor with the lowest cost and transferred its reference testing to this new vendor. Subsequently, the hospital laboratory medical director observed some shortcomings in the clinical performance of a number of test methods.

### Explanation and Consequences

The hospital laboratory management team initially evaluated the various vendors' proposals solely on cost. Other factors, such as clinical performance, value-added services, and clients' references, were not considered. The problems with the vendor's performance impacted the clinicians' ability to treat patients. Thus, selecting the vendor solely based on low cost did cause the laboratory problems with dissatisfied clinicians.

In accordance with the hospital–vendor contract, formal notification of performance failures was then regularly reported to the vendor. Over time, the vendor's performance failed to meet standards and per contract, the business relationship was terminated pending the required 90-day notification period. During this notification period, laboratory management engaged in contract discussions with another vendor who had participated in the competitive bid process. In time, the laboratory was able to engage a reference testing service that met key performance requirements.

# STANDARDS OF PERFORMANCE

▩ Even when a contract is in place with a vendor, the cost for reference testing can spiral out of control when providers are given carte blanche to both order tests and select the reference testing site. Laboratory leadership must monitor "leakage" of tests to noncontracted vendors and where appropriate, engage the clinicians, and redirect the tests.

▩ Common business practice does require a competitive bid for those services that incur significant expense. However, the selection of a reference laboratory service should be evaluated on clinical performance and service requirements as well as cost. Consideration of these three elements can better assure that the necessary diagnostic needs for patient care are addressed in addition to the business need to effectively control costs.

# Instrument Selection for the Clinical Laboratory

## OVERVIEW

The diagnostic laboratory industry is composed of manufacturers who compete in both the national and global marketplace. There are a large number of instruments available to meet the needs of both commercial and hospital-based laboratories of all sizes. It is imperative to clearly understand patient care needs in order to properly select instruments. Selection criteria should include clinical method evaluation, reagent stability, throughput capacity, ease of use and process control, and documented instrument performance and vendor service record.

## UNDERSTAND PATIENT CARE NEEDS

The acquisition of any major instrument requires a thorough consideration of both current patient care needs and any new programs. Failure to do so may create a situation where the laboratory is unable to properly support patient care needs over the life of the instrument.

### Case with Error

A centralized laboratory supported a multihospital system and the laboratory was engaged in the selection of a new primary chemistry analyzer and automated specimen transport line. This integrated system was expected to have a useful life of 5 to 7 years. The hospital system was constructing a pediatric hospital that would open for patient care about 5 years after the new integrated transport and instrument system would be installed. It was expected that the centralized laboratory would provide testing to the new pediatric facility.

The analyzer and specimen transport system were selected and installed and adequately supported patient testing needs at the time of installation. However, with the opening of the pediatric hospital, numerous problems were encountered in processing microtainer samples on the integrated system.

### Explanation and Consequences

The laboratory management team had appropriately defined patient testing needs based on its current volume at the time, which were more than 90% adult specimens. However, a number of external circumstances had delayed the implementation of the new chemistry analyzer and automated specimen transport line. By the time it was installed and operational, the pediatric hospital was open for service at the beginning of the useful life for the integrated system, not near the end of its useful life. With the opening of the pediatric hospital, the laboratory experienced a shift in the text mix so that now 25% of the total test volume was pediatric specimens and the vast majority of these were collected in microtainers.

To properly support this patient population, the laboratory established a pediatric test bench with a chemistry analyzer that could accommodate the smaller specimen volumes.

# EVALUATE COMPETITIVE PRODUCTS

▶ Many laboratory instruments are offered by manufacturers who compete in the global marketplace. As in any industry, these competitive forces drive manufacturers to continuously enhance their instruments and reagents. Laboratory management should engage in a competitive evaluation of products to assure that it is providing the latest technology that best meets patient care needs and optimizes the efficient use of resources.

## *Case with Error*

For years, a laboratory utilized the same manufacturer for its coagulation instrument and reagents. The products performed acceptably. However, it had been more than a decade since any other instrument and reagent system had been evaluated. Thus, there had not been an objective assessment of this product's performance capabilities and weaknesses.

## *Explanation and Consequences*

A failure to periodically evaluate current technology in the marketplace can impact patient care as well as the laboratory's operations and finances. Over time, there are shifts in the patient diagnoses and acuity mix, which may alter the performance requirements for a test method. A competitive assessment of several products can identify technology enhancements that can support optimum patient care. It also encourages lower pricing since the vendors must compete on both performance and price in order to secure the contract.

In this particular situation, new laboratory leadership required an evaluation of a minimum of three vendors. The outcome of this competitive assessment allowed for the selection of new instrument technology with greater throughput and at a much lower cost.

## Case with Error Averted

A laboratory was evaluating two competing vendors for new technology in molecular diagnostic testing. Both vendors had a positive reputation in the field. There were a few "early adapter" laboratories that had implemented the new technology, and the users were evenly divided between the two vendors. The laboratory conducted a thorough assessment of both instrument systems. The clinical outcomes were statistically equivalent. However, the preanalytic preparation technique for one method required significantly more manual manipulation when compared with the other method. The laboratory selected the instrument technology that had a more streamlined process for specimen preparation. Within the next 2 years, this instrument technology became the dominant method for this test.

## Explanation and Consequences

When the laboratory team conducted its evaluations for these products, it had determined that there were two drawbacks with the specimen preparation method for the more complicated manual method. Since the specimen preparation was a manual method with a number of steps, that process inherently posed a greater rate for introducing error that could affect results. It also required more "hands-on" technologist time which meant that as volume increased, more staffing would be required at an earlier point when compared to the competing technology.

These performance concerns were valid as, over time, one instrument came to dominate the market. These market forces also motivated the other vendor to invest in research and develop an improved specimen preparation method.

## Case with Error

A laboratory was conducting a comparative assessment of new instruments. The clinical methodologies were comparable, service needs such as throughput capacity and TAT were fairly equivalent, and all vendors had solid performance track records. It was decided to select the vendor with the lowest cost per test. Post-implementation, a performance review was conducted and while the instrument's clinical performance met all requirements, its actual cost per test was higher than expected.

## Explanation and Consequences

During its evaluation, the management team had overlooked costs associated with quality control and instrument maintenance. It is important to gain a complete picture of all costs related to an instrument and test method. The comparative assessment should include these costs for all vendors. In this situation, even if the final outcome may still have been the selection of the same instrument, the management team could have leveraged the competitive bid process to secure better pricing.

# STANDARDS OF PERFORMANCE

▓ The acquisition of new instrument technology must consider the current and future patient care needs. Laboratory managers should collaborate with key departments, such as finance, business development, or marketing, to identify programs planned for future implementation that will require support from the new instrument. Clearly defining current and anticipated patient testing needs can improve the instrument selection process, provide for discounted costs triggered by a volume targets, and enhanced performance requirements from the vendor.

▓ When an instrument has reached the end of its useful life, this presents an opportunity to comparatively assess new technology and select an instrument and reagent system that can best support patient care. Many instruments have a useful life of 5 to 7 years or sometimes more. However, the competitive forces in the marketplace drive technological innovations during that same period of time. These innovations can benefit patient care and laboratory operations and should be evaluated.

▓ Clinical performance is crucial to evaluating instruments. However, many other factors must be considered as well. Methods should be assessed to assure that staff can work efficiently and in a manner that minimizes the risk of errors. The financial assessment should identify "hidden" costs that may be associated with quality control, instrument maintenance, or other factors such as renovation expense that may be required for installation.

# Selection and Utilization of a Laboratory Information System

## OVERVIEW

The laboratory information system (LIS) represents a strategic investment in data management capabilities and capital funds. These laboratory test results provide essential information to providers for the diagnosis and treatment of patients in both inpatient and outpatient settings. Therefore, the selection and ongoing operation of an LIS requires participation from a broad array of users, clients, and key stakeholders.

When selecting an LIS, the users, both pathologists and technologists, must identify performance requirements such as capabilities for managing and processing test orders, interfacing with various instruments and other software information systems, manipulation of data to support numerous requirements for results reporting, and supporting ancillary functions such as billing, client services and quality management, and process control. A dedicated team from across the clinical and anatomical pathology services should define and rank the performance requirements. These criteria can serve as an objective tool to evaluate each LIS application.

Laboratory clients include physicians and nurses. They may have test order or result report needs for patient care that should be considered when assessing LIS applications.

Key stakeholders include other departments that must interact on some level with the LIS. For example, the information technology group provides services such as network support or interface design and maintenance associated with other software applications. Finance is another department that may rely on the LIS for obtaining coding data and completing claims submission. It is necessary to engage these key stakeholders and assure that the LIS will not compromise any necessary operational processes with external departments.

Lastly, the management team must understand the level of staff expertise that is required to support the LIS, including adequate staff coverage and the necessary training of users.

## DETERMINE STAFF SUPPORT FOR LIS

▶ The laboratory leadership should manage the LIS installation so that its performance is optimized to support data management. When extensive custom modifications are required for the software, management should understand and plan for the necessary support resources.

### Case with Error

A tertiary medical center installed one of the most commonly used LIS applications. However, a number of significant software modifications were made in an effort to provide some unique services. In accordance with the vendor's guidelines developed for the routine software application, the recommended number of staff was trained to provide routine support and maintenance. However, the laboratory experienced extended system downtimes when performing "routine" software upgrades. In some instances, these downtimes were as long as 1 to 2 days.

### Explanation and Consequences

A particular LIS application had a solid performance record within the industry. An institution of similar size purchased and installed the same product with minimal customization and routinely handled software upgrades with minimal interruption to service. In this circumstance, the highly customized software presented unique challenges that made it difficult to easily migrate to new releases. This problem was further exacerbated as there was staff turnover within the vendor-trained members of the hospital's LIS support team. Given the extensive customizations to the software, the laboratory should have pursued a contractual agreement for more extensive vendor service or configured its on-site LIS support staff with software analysts who were more extensively trained.

## CONFIRM COMPATABILITY WITH OTHER SOFTWARE APPLICATIONS

> There are a number of software applications that must be interfaced to an LIS. The "owners" of these external information systems should participate in evaluating the new LIS products to assure that there is an acceptable degree of compatibility. Failure to do so can create unnecessary problems.

### Case with Error

A hospital laboratory's selection team identified its preferred product to replace its existing LIS based on various performance capabilities, including its ability to support testing in the outreach market. The hospital and laboratory leadership team had planned to increase its activity in this market. After a new LIS was implemented, the laboratory encountered significant problems with managing test orders placed for the inpatient population via an external computerized order entry system.

### Explanation and Consequences

The outreach market is primarily focused on providing test results in an outpatient setting and does not have the need for providing tests performed at regular intervals as in the acute inpatient care setting. To resolve this problem, the laboratory had to incur the unplanned cost to develop, install, and maintain middleware.

### Case with Error

An extensive database built for a new LIS was completed. Post-implementation, it was determined that the LIS billing function was not able to send modifiers for CPT codes across the interface to the institution's billing and claims submission software application.

### Explanation and Consequences

When building the database for the new LIS, the programming requirements necessary for successful integration with other software applications must be considered. In this case, there had been a failure to thoroughly vet the billing function. This matter required a significant investment of both staff and financial resources to correct this problem and assure that full reimbursement for testing could be obtained.

# SUPPORT CLIENTS' SERVICE EXPECTATIONS

▶ Periodically, a laboratory will expand its test menu. When providing a new service, it is important to understand the service requirements that clients may have for these result reports.

## *Case with Error Averted*

A laboratory was adding cytogenetics as a new testing service. As part of the implementation planning, the laboratory team designed several different report formats, all of which could be supported by the LIS. It contacted key clinicians and solicited their evaluation of the report formats. The clients identified data elements and report formatting needs that were missing from the "standard" report options in the LIS. It was determined that the LIS could support these needs with some nominal software modifications.

## *Explanation and Consequences*

By engaging the clients, the laboratory management team was able to enhance the LIS report so that it could better meet the clinicians' needs. The laboratory team had determined that the standard report formats in the LIS met all regulatory standards for reporting cytogenetics results.

# STANDARDS OF PERFORMANCE

▩ It is the responsibility of laboratory leadership to understand and define the resources required to support the LIS. Resource needs can vary depending on the complexity of the LIS. A highly customized LIS will require staff with programming expertise, whereas a simple "turnkey" application will require vendor-trained staff.

▩ The LIS provides critical support for data management of test orders and results. Given the laboratory's key role in patient diagnosis and treatment, it is absolutely essential that the LIS can effectively communicate with various independent software applications. The laboratory leadership is responsible for engaging key stakeholders of external software applications and assuring that all necessary performance requirements can be met.

▩ When selecting a new LIS, a laboratory team should be formed with representation from across numerous subspecialties. The team should define and prioritize performance requirements that can then be used as an objective tool to measure capabilities of various LIS applications.

▩ Clients' needs for test ordering or result reporting should be solicited. They should be engaged when evaluating a new LIS and also when there is a service update.

# Suggested Reading

A number of textbooks are available that can provide a more extensive discussion of management concepts and practices that are applicable to the laboratory setting. The following is a brief list of resources for the interested reader:

Garcia LS, ed. Clinical Laboratory Management. Washington, DC: ASM Press; 2004.

Harmening D. Laboratory Management: Principles and Processes. 2nd ed. Philadelphia, PA:FA Davis Co.; 2007.

Hudson J. Principles of Clinical Laboratory Management: A Study Guide and Workbook. 1st ed. Upper Saddle River, NJ: Prentice Hall; 2003.

Lewandrowski K, ed. Clinical Chemistry Laboratory Management and Clinical Correlations. 1st ed. Philadelphia, PA: Lippincott Williams & Wilkins; 2002.

O'Brien JA. Common Problems in Clinical Laboratory Management. New York, NY: McGraw-Hill Co.; 1999.

Varnadoe LA. Medical Laboratory Management and Supervision: Operations, Review and Study Guide. Philadelphia, PA:FA Davis Co.; 1996.

# Index